Alfred Alcock

# Illustrations of the Zoology of H.M. Indian Marine Surveying

## Steamer Investigator

Under the command of Commander A. Carpenter, R.N., D.S.O., and of

Commander R.F. Hoskyns, R.N.

Alfred Alcock

**Illustrations of the Zoology of H.M. Indian Marine Surveying Steamer Investigator**
*Under the command of Commander A. Carpenter, R.N., D.S.O., and of Commander R.F. Hoskyns, R.N.*

ISBN/EAN: 9783337384364

Printed in Europe, USA, Canada, Australia, Japan

Cover: Foto ©berggeist007 / pixelio.de

More available books at **www.hansebooks.com**

# ILLUSTRATIONS

OF THE

# ZOOLOGY

OF THE

## ROYAL INDIAN MARINE SURVEY SHIP

# INVESTIGATOR

UNDER THE COMMAND OF

### COMMANDER W. G. BEAUCHAMP, R.I.M.

---

### MOLLUSCA—Pt. VI, Pls. XXI—XXIII.

UNDER THE DIRECTION OF

N. ANNANDALE, D.Sc., C.M.Z.S., F.L.S., SUPERINTENDENT OF THE INDIAN MUSEUM,

AND OF

F. H. STEWART, M.A., D.Sc., M.B., SURGEON-NATURALIST TO THE INDIAN MARINE SURVEY.

Published under the Direction of Captain W. Lumsden, R.A.,
Director of the Royal Indian Marine.

CALCUTTA
SUPERINTENDENT GOVERNMENT PRINTING, INDIA
1909

*Price One Rupee Eight Annas.*

# EXPLANATION OF PLATES.

## MOLLUSCA.

### PLATE I.

*Figs. 1, 1a, 1b, 1c.—Nuculana indica*, E. A. Smith, Annals and Magazine of Natural History, (6) XVI. July, 1895, p. 16.

*Figs. 2, 2a, 2b, 2c.—Nuculana fumosa*, E. A. Smith, Annals and Magazine of Natural History, (6) XVI. July, 1895, p. 16.

*Figs. 3, 3a, 3b, 3c.—Nucula (Acila) fultoni*, E. A. Smith.

*Figs. 4, 4a, 4b, 4c.—Nucula bengalensis*, E. A. Smith, Annals and Magazine of Natural History, (6) XVI. July, 1895, p. 15.

*Figs. 5, 5a, 5b, 5c.—Nucula donaciformis*, E. A. Smith, Annals and Magazine of Natural History, (6) XVI. July, 1895, p. 15.

1c. ×5    1b. ×3    1a. ×3    1. ×3

2c. ×5    2b. ×4    2a. ×4    2. ×4

3c. ×2    3b.    3a.    3.

4c. ×4    4b. ×2    4a. ×2    4. ×2

5c. ×4    5b. ×2    5a. ×2    5. ×2

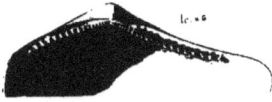

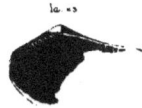

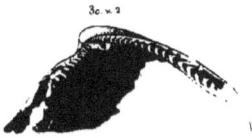

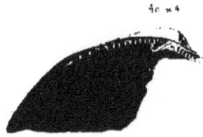

1. Nuculana indica, E. A. Smith    2. Nuculana fumosa, E. A. Smith    3. Nucula fultoni, E. A. Smith
4. Nucula bengalensis, E. A. Smith    5. Nucula donaciformis, E. A. Smith

S. C. Mondul, del.    Photoetching Survey of India Offices, Calcutta

# EXPLANATION OF PLATES.

## MOLLUSCA.

### PLATE II.

*Figs. 1, 1a.—Amussium andamanicum*, E. A. Smith, Annals and Magazine of Natural History, (6) XIV. September, 1894, p. 172 ; and (6) XVI. September, 1895, p. 265.

*Figs. 2, 2a.—Amussium solitarium*, E. A. Smith, Annals and Magazine of Natural History, (6) XIV. September, 1894, p. 173.

*Figs. 3, 3a.—Amussium alcocki*, E. A. Smith, Annals and Magazine of Natural History, (6) XIV. September, 1894, p. 172.

*Figs. 4, 4a.—Lucina bengalensis*, E. A. Smith, Annals and Magazine of Natural History, (6) XIV. September, 1894, p. 171.

*Figs. 5, 5a.—Glauconome sculpta*, Sowerby, Proceedings of the Malacological Society, Vol. I. March, 1894, p. 40.

*Figs. 6, 6a, 6b, 6c.—Yoldia angulata*, Sowerby.

*Figs. 7, 7a, 7b.—Malletia conspicua*, E. A. Smith, Annals and Magazine of Natural History, (6) XVI. July, 1895, p. 17.

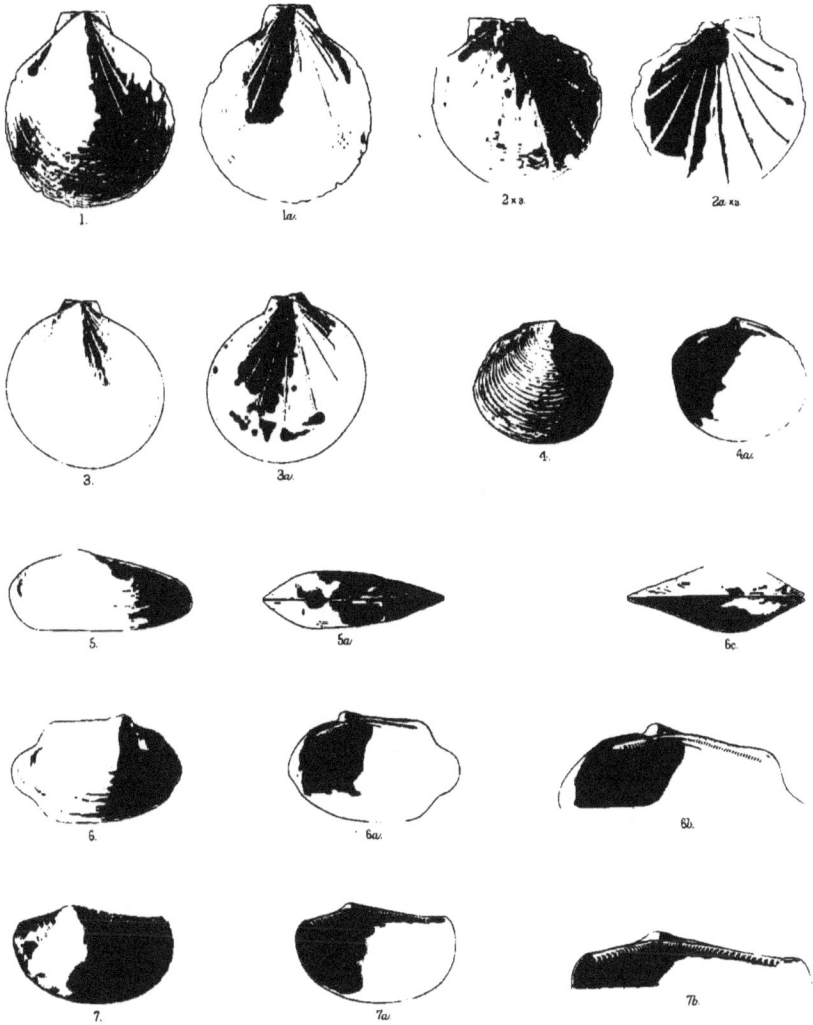

1. Amussium andamanicum. E. A. Smith. 2. Amussium solitarium. E. A. Smith. 3. Amussium alcocki. E. A. Smith.
4. Lucina bengalensis. E. A. Smith. 5. Glauconome sculpta. Sowb. 6. Yoldia angulata. Sowb.
7. Malletia conspicua. E. A. Smith.

S. C. Mondul del.

Photo-etching-Survey of India Offices, Calcutta, July 1897

# EXPLANATION OF PLATES.

## MOLLUSCA.

### PLATE III.

*Figs.* **1, 1a, 1b.**—*Cryptodon acuticarinatus*, E. A. Smith, Annals and Magazine of Natural History, (6) XVI. July, 1895, p. 14.

*Figs.* **2, 2a, 2b.**—*Cryptodon investigatoris*, E. A. Smith, Annals and Magazine of Natural History, (6) XVI. July, 1895, p. 13.

*Figs.* **3, 3a, 3b.**—*Lucina spinifera*, (Montagu) E. A. Smith, Annals and Magazine of Natural History, (6) XVI. p. 12.

*Figs.* **4, 4a, 4b, 4c.**—*Limopsis indica*, E. A. Smith, Annals and Magazine of Natural History, (6) XIV. September, 1894, p. 171.

*Figs.* **5, 5a.**—*Cuspidaria macrorhynchus*, E. A. Smith, Annals and Magazine of Natural History, (6) XVI. July, 1895, p. 12.

*Fig.* **6.**—*Cuspidaria (Cardiomya) alcocki*, E. A. Smith, Annals and Magazine of Natural History, (6) XIV. September, 1894, p. 170.

*Figs.* **7, 7a.**—\**Abra maxima*, Sowerby, Proceedings Malacological Society, 1894, Vol. I. p. 40, pl. v. fig. 5, E. A. Smith, Annals and Magazine of Natural History, (6) XIV. 1894, p. 169; and XVI. 1895, p. 10.

*Figs.* **8, 8a.**—*Abra convexior*, E. A. Smith, Annals and Magazine of Natural History, (6) XVI. July, 1895, p. 10.

*Figs.* **9, 9a.**—*Tellina parvula*, E. A. Smith, Annals and Magazine of Natural History, (6) XVI. July, 1895, p. 10.

---

\* *Syndesmya maxima.*

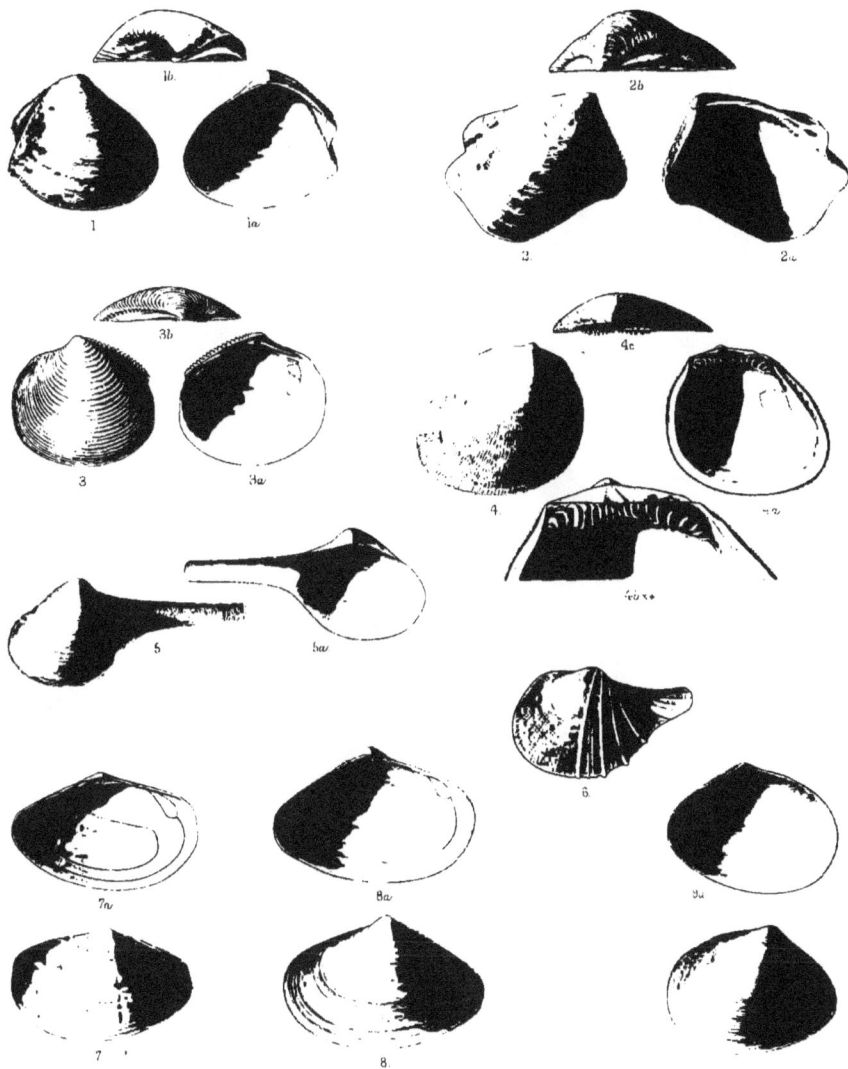

# EXPLANATION OF PLATES.

## MOLLUSCA.

### PLATE IV.

*Figs. 1a, 1b, 1c, 1d, 1e.—Verticordia (Euciroa) eburnea,* Wood-Mason and Alcock, Annals and Magazine of Natural History, (6) VIII. December, 1891, p. 447.

*Figs. 2, 2a.—Lima (Limatula) subtilis,* E. A. Smith, Annals and Magazine of Natural History, (6) XVI. July, 1895, p. 18.

*Figs. 3, 3a, 3b.—Crassatella indica,* E. A. Smith, Annals and Magazine of Natural History, (6) XVI. September, 1895, p. 265.

*Figs. 4, 4a.—Venus juvenilis,* E. A. Smith, Annals and Magazine of Natural History, (6) XVI. July, 1895, p. 9.

*Figs. 5, 5a.—Cytherea (Caryatis) pudicissima,* E. A. Smith, Annals and Magazine of Natural History, (6) XIV. September, 1894, p. 169.

*Figs. 6, 6a.—Cryptodon philippinarum, (Hanley),* E. A. Smith, Annals and Magazine of Natural History, (6) XVI, p. 13.

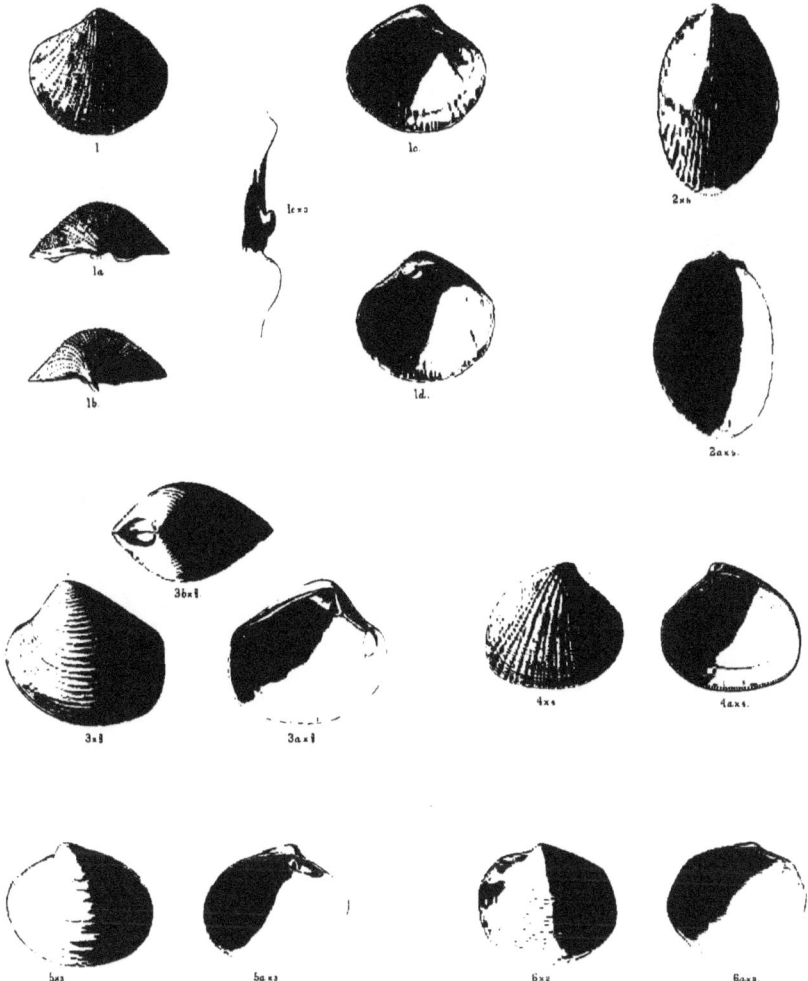

1. Verticordia (Euciroa) eburnea, W. M. & A. 1e x 2.  2. Luna subtilis, E. A. Smith x 5.  3. Crassatella indica, E. A. Smith x 4.
4. Venus juvenilis, E. A. Smith x 4.  5. Cytherea (Caryatis) pudicissima, E. A. Smith x 3.  6. Cryptodon philippinarum, Hanley x 2.

S. C. Mondul, del.

Photo-etching; Survey of India Offices, Calcutta, August 1897

EXPLANATION OF PLATES.

———+———

# MOLLUSCA.

———+———

## PLATE V.

*Figs. 1, 1a.—Pontiothauma mirabile*, E. A. Smith, Annals and Magazine of Natural History, (6), XVI. July, 1895, p. 2.

*Figs. 2, 2a.—Pontiothauma abyssicola*, E. A. Smith, Annals and Magazine of Natural History, (6), XIV. July, 1895, p. 3.

1

1a

2

2a

1  Pontiothauma mirabile   E. A. Smith

2  Pontiothauma abyssicola.   E. A. Smith

.

# EXPLANATION OF PLATES.

## MOLLUSCA.

### PLATE VI.

*Figs. 1, 1a, 1b.*—*Murex malabaricus*, E. A. Smith, Annals and Magazine of Natural History, (6), XIV. September, 1894, p. 162.

*Figs. 2, 2a.*—*Pirula investigatoris*, E. A. Smith, Annals and Magazine of Natural History, (6), XIV. November, 1894, p. 367.

1.

1a.

1b

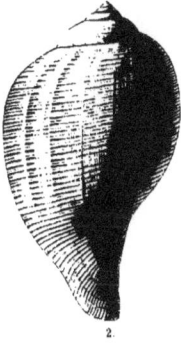

2.

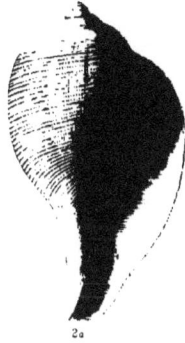

2a.

Matt. – Pt II, Pt, VII – VIII
1898.

# MOLLUSCA.

## PLATE VII.

*Figs. 1, 1a.—Pleurotoma (Ancistrosyrinx) travancorica,* E. A. Smith, Annals and Magazine of Natural History, (6) XVIII, 1896, p. 368.

*Figs. 2, 2a.—Pleurotoma (Surcula) profundorum,* E. A. Smith, Annals and Magazine of Natural History, (6) XVIII, 1896, p. 369.

*Figs. 3, 3a.—Natica (Lunatia) abyssicola,* E. A. Smith, Annals and Magazine of Natural History, (6) XVIII, 1896, p. 370.

*Figs. 4, 4a.—Natica (Lunatia) levis,* E. A. Smith, Annals and Magazine of Natural History, (6) XVIII, 1896, p. 370.

*Figs. 5, 5a.—Dentalium magnificum,* E. A. Smith, Annals and Magazine of Natural History, (6) XVIII, 1896, p. 371.

*Figs. 6, 6a.—Cardium (Fragum) simillimum,* E. A. Smith, Annals and Magazine of Natural History, (6) XVIII, 1896, p. 372.

1.2                1a.2                                    2.2              2a.2.

3.3                3a..3                                   4.2.             4a.2.

5.

6.2.

5a.

6a.2.

1, Pleurotoma (Ancistrosyrinx) travancorica, E A Smith.    2, Pleurotoma (Surcula) profundorum, E A Smith.
3, Natica (Lunatia) abyssicola, E A Smith.    4, Natica (Lunatia) lævis, E A Smith.
5, Dentalium magnificum, E A Smith    6, Cardium Fragum simillimum, E. A. Smith

S C Moudul, del
Photo-etching                                                              Survey of India Offices, Calcutta, August, 1898

# EXPLANATION OF PLATES.

## MOLLUSCA.

### PLATE VIII.

*Figs. 1, 1a, 1b, 1c.*—*Yoldia anatina*, E. A. Smith, Annals and Magazine of Natural History, (6) XVIII, 1896, p. 373.

*Figs. 2, 2a.*—*Cuspidaria approximata*, E. A. Smith, Annals and Magazine of Natural History, (6) XVIII, 1896, p. 373.

*Figs. 3, 3a.*—*Myonera bicarinata*, E. A. Smith, Annals and Magazine of Natural History, (6) XVIII, 1896, p. 374.

*Figs. 4, 4a.*—*Scrobicularia ceylonica*, E. A. Smith, Annals and Magazine of Natural History, (6) XVIII, 1896, p. 375.

...b                                        ...

1b × 3                                    1; × 3

2 × 3.        2 × 3         3 × 3        3. × 3

4 × 2                        4. × 2

1  Yoldia anatina, E A Smith      2   Cuspidaria approximata, E A Smith
3  Myonera bicarinata, E A Smith      4   Scrobicularia ceylonica, E A Smith

*1901*

# EXPLANATION OF PLATE.

## MOLLUSCA.

### PLATE IX.

*Figs. 1, 1a.*—*Pleurotoma optata*, E. A. Smith, Annals and Magazine of Natural History, series 7, Vol. IV. October, 1899, p. 238.

*Figs. 2, 2a.*—*Pleurotoma oldhami*, E. A. Smith, *loc. admodum cit.*

*Figs. 3, 3a.*—*Pleurotoma (Surcula) breviplicata*, E. A. Smith, *loc. admodum cit.*

*Figs. 4, 4a.*—*Pleurotoma (Surcula) eurina*, E. A. Smith, *tom. cit.*, p. 239.

*Figs. 5, 5a.*—*Pleurotoma (Surcula) praecipua*, E. A. Smith, *loc. admodum cit.*

*Figs. 6, 6a.*—*Pleurotoma (Surcula) arcana*, E. A. Smith, *loc. admodum cit.*

1×2       1a×2

2×1½      2a×1½

3×2½      3a×2½

4×1½      4a×1½

5×2       5a×2

6×3       6a×3

1. Pleurotoma optata.   2. Pleurotoma oldhami.   3. Pleurotoma (surcula) brevifienta.   4. Pleurotoma (surcula) eurina.
5. Pleurotoma (surcula) praecipua.   6. Pleurotoma (surcula) arcana.

S. C. Mondul, del.

Photogravure, Survey of India Offices, Calcutta, August 1901.

# EXPLANATION OF PLATE.

# MOLLUSCA.

### PLATE X.

*Figs. 1, 1a.*—*Drillia investigatoris*, E. A. Smith, Annals and Magazine of Natural History, series 7, Vol. IV. October, 1899, p. 240.

*Fig. 2.*—*Drillia captiva*, E. A. Smith, *loc. admodum cit.*

*Fig. 3.*—*Drillia capta*, E. A. Smith, *loc. admodum cit.*

*Figs. 4, 4a.*—*Trophon tenuirostratus*, E. A. Smith, *tom. cit.*, p. 241.

*Figs. 5, 5a.*—*Trophon indicus*, E. A. Smith, *loc. admodum cit.*

*Figs. 6, 6a.*—*Nassaria laevior*, E. A. Smith, *tom. cit.*, p. 242.

*Figs. 7, 7a.*—*Tritonidea delicata*, E. A. Smith, *loc. admodum cit.*

*Figs. 8, 8a.*—*Fusus captivus*, E. A. Smith, *loc. admodum cit.*

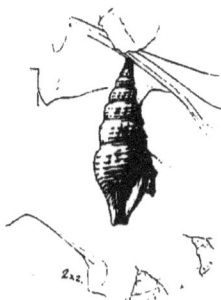

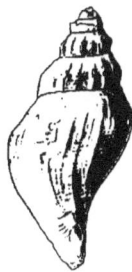

1. Drilla investigatoris. 2. Drilla captiva. 3. Drilla captiva. 4. Trophon tenuirostratus. 5. Trophon indicus.
6. Nassaria laevior. 7. Tritonidea delicata. 8. Fusus captivus.

# EXPLANATION OF PLATE.

## MOLLUSCA.

### PLATE XI.

*Figs. 1, 1a.*—*Pisania angusta*, E. A. Smith, Annals and Magazine of Natural History, series 7, Vol. IV, October, 1899, p. 243.

*Figs. 2, 2a.*—*Nassa aracanensis*, E. A. Smith, *loc. admodum cit.*

*Figs. 3, 3a.*—*Nassa diluta*, E. A. Smith, *loc. admodum cit.*

*Figs. 4, 4a.*—*Ancilla leucospira*, E. A. Smith, *tom. cit.*, p. 245.

*Figs. 5, 5a.*—*Cancellaria cretacea*, E. A. Smith, *loc. admodum cit.*

*Figs. 6, 6a.*—*Ancilla glans*, E. A. Smith, *tom. cit.*, p. 246.

*Figs. 7, 7a.*—*Columbella (Mitrella) supraplicata*, E. A. Smith, *tom. cit.*, p. 244.

*Figs. 8, 8a.*—*Coralliophila indica*, E. A. Smith, *loc. admodum cit.*

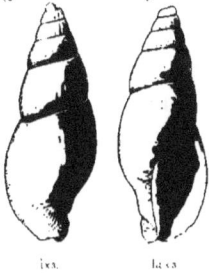

1×3. 1a×3

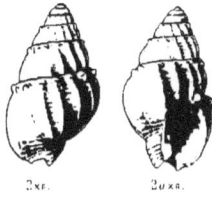

2×5. 2a×5.

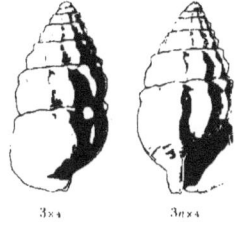

3×4 3a×4

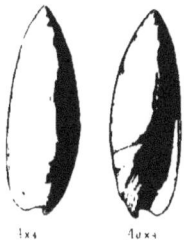

4×4 4a×4

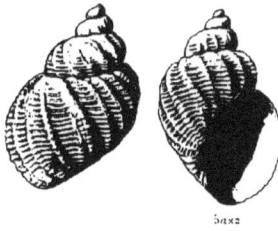

5a×2

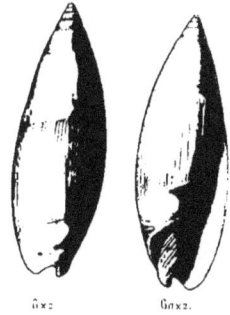

6×2 6a×2.

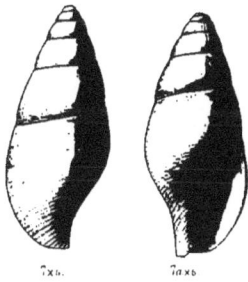

7×6. 7a×6

8×2½ 8a×2½

# EXPLANATION OF PLATE.

## MOLLUSCA.

### PLATE XII.

*Figs. 1, 1a.—Scalaria bengalensis*, E. A. Smith, Annals and Magazine of Natural History, series 7, Vol. IV, October, 1899, p. 246.

*Figs. 2, 2a.—Scalaria subcasta*, E. A. Smith, *loc. admodum cit.*

*Figs. 3, 3a.—Leptothyra delecta*, E. A. Smith, *tom. cit.*, p. 248.

*Figs. 4, 4a-c.—Astralium bathyraphe*, E. A. Smith, *tom. cit.*, p. 247.

*Figs. 5, 5a.—Turbo incoloratus*, E. A. Smith, *loc. admodum cit.*

*Figs. 6, 6a-b.—Solariella oxycona*, E. A. Smith, *tom. cit.*, p. 248.

*Figs. 7, 7a.—Puncturella (Cranopsis) indica*, E. A. Smith, *tom. cit.*, p. 249.

*Figs. 8, 8a.—Fissurella delicata*, E. A. Smith, *loc. admodum cit.*

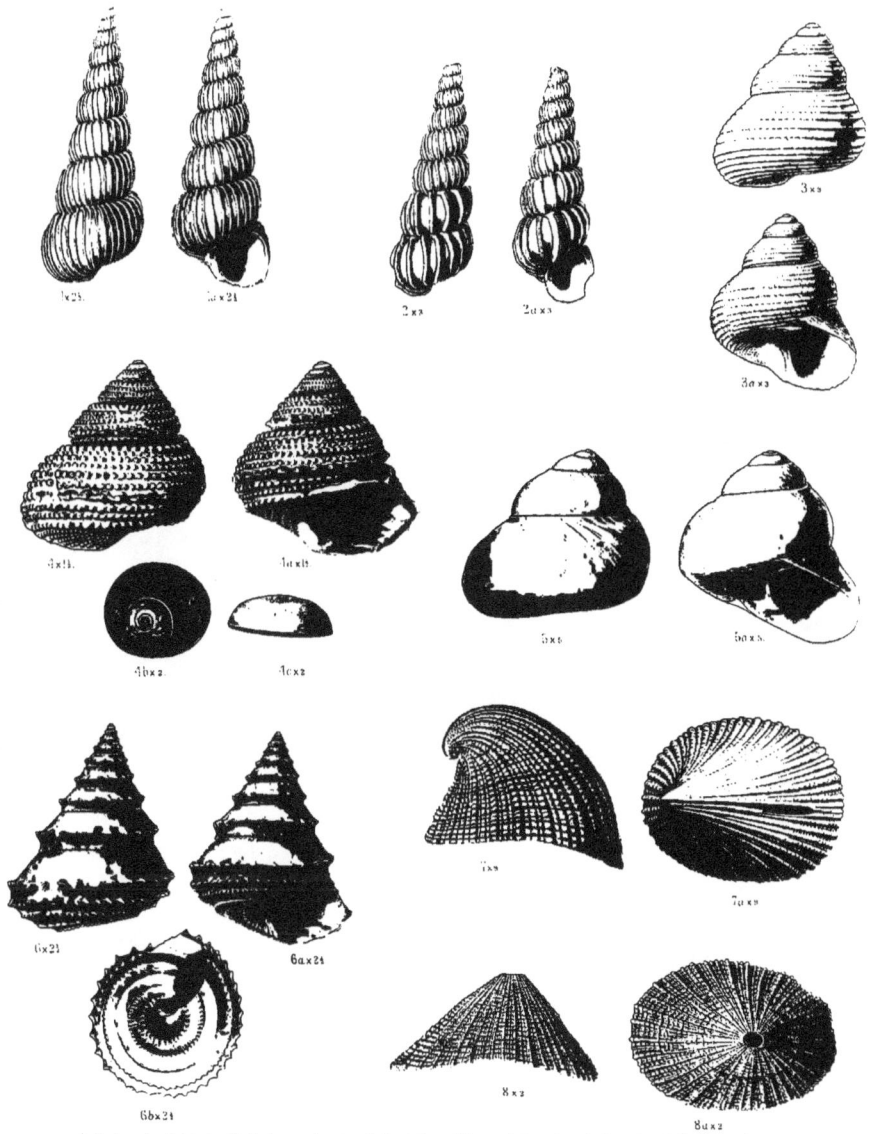

1x2.                    1a x 2½                    2 x 3        2a x 3        3 x 3

3a x 3

4 x 3.        4a x 3.

4b x 2        4c x 2

5 x 3        6a x 3.

6 x 2½        6a x 2½        7 x 3

7a x 3

6b x 2½        8 x 2        8a x 2

1. Scalaria bengalensis  2. Scalaria subcasta  3. Leptothyra delecta  4. Astralium bathyraphe.  5. Turbo incoloratus
6. Solariella oxycona.  7. Puncturella (Cranopsis) indica.  8. Fissurella delicata.

EXPLANATION OF PLATE.

# MOLLUSCA.

### PLATE XIII.

*Figs. 1, 1a.—Tellina travancorica*, E. A. Smith, Annals and Magazine of Natural History, series 7, Vol. IV, October, 1899, p. 249.

*Figs. 2, 2a.—Abra affinis*, E. A. Smith, *tom. cit.*, p. 250.

*Figs. 3, 3a.—Arca (Barbatia) incerta*, E. A. Smith, *tom. cit.*, p. 251.

*Figs. 4, 4a.—Lima indica*, E. A. Smith, *loc. admodum cit.*

*Figs. 5, 5a.—Myodora quadrata*, E. A. Smith, *tom. cit.*, p. 250.

1×1½

2×2½

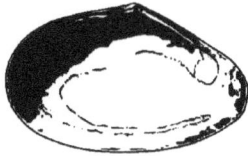

1a×1½

2a×2½

4a.

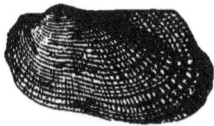

3×2½

4

5×4

3a×2½

5a×4

1  Tellina travancorica   2  Abra affinis   3  Arca (Barbatia) incerta   4  Lima indica   5  Myodora quadrata

C. Mondul, del.

Photogravure, Survey of India Offices, Calcutta, July 1905.

MaP. — ??: IV, ??? XIV–XVI
1907

# EXPLANATION OF PLATE.

## MOLLUSCA.

### PLATE XIV.

*Figs. 1, 1a.*—*Conus semisulcatus*, Sowerby, Proc. Zool. Soc., 1870, p. 257; and E. A. Smith, Ann. Mag. Nat. Hist., Ser. 7, Vol. XIII, June, 1904, p. 454.

*Figs. 2, 2a.*—*Pleurotoma (Surcula) margaritæ*, E. A. Smith, Ann. Mag. Nat. Hist., Ser. 7, Vol. XIII, June, 1904, p. 458.

*Figs. 3, 3a.*—*Pleurotoma vagata*, E. A. Smith, *Op. cit.*, p. 456.

*Figs. 4, 4a.*—*Drillia worthingtoni*, E. A. Smith, *Op. cit.*, p. 460.

1.    1a.

2.    2a.

3a.

3.    4a.

Conus semisulcatus, Sow. 2. Pleurotoma (Surcula) Margaritae, E. A. Smith
3. Pleurotoma vagata, E. A. Smith. 4. Drilia Worthingtoni, E. A. Smith.

# EXPLANATION OF PLATE.

## MOLLUSCA.

### PLATE XV.

*Fig. 1.—Latiaxis diadema* (A. Adams). E. A. Smith, Ann. Mag. Nat. Hist., Ser. 7, Vol. XIII, June, 1904, p. 466.

*Murex diadema*, A. Adams, Proc. Zool. Soc., 1853, p. 70.

*Figs. 2, 2a.—Clathurella perlissa*, E. A. Smith, Ann. Mag. Nat. Hist., Ser. 7, Vol. XIII, June, 1904, p. 460.

*Figs. 3, 3a.—Marginella grisea* (Jousseaume). E. A. Smith, Ann. Mag. Nat. Hist., Ser. 7, Vol. XIII, June, 1904, p. 468.

*Persicula grisea*, Jousseaume, Rev. Mag. Zool., 1875, p. 268.

*Figs. 4, 4a.—Lacuna globosa*, E. A. Smith, Ann. Mag. Nat. Hist., Ser. 7, Vol. XIII, June, 1904, p. 472.

*Figs. 5, 5a.—Chiton ceylanicus*, E. A. Smith, Ann. Mag. Nat. Hist., Ser. 7, Vol. XIV, July, 1904, p. 7.

1.

2.

2a.

3

3a.

4

4a.

5

5a.

1. Latiaxis diadema A Adams. × 3. 2. Clathurella perlissa E A Smith. × 5. 3. Marginella grisea Jouss × 24. 4. Lacuna globosa. E.A Smith × 10.
5. Chiton ceylanicus. E.A Smith. × 4

# MOLLUSCA.

### PLATE XVI.

*Figs. 1, 1a, 1b.—Capulus fragilis*, E. A. Smith, Ann. Mag. Nat. Hist., Ser. 7, Vol. XIV, July, 1904, p. 1.

*Figs. 2, 2a.—Scaphander ceylanica*, E. A. Smith, *Op. cit.*, p. 5.

*Figs. 3, 3a, 3b.—Acmæa minutissima*, E. A. Smith, *Op. cit.*, p. 4.

1a

1                    1b

2                    2a.

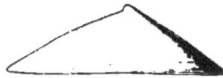

3b.

3                                        3b.

1  Capulus fragilis, E A Smith, × 4     2  Scaphander ceylanica, E A Smith, × 6
3  Acmaea minutissima, E A Smith, × 16

# EXPLANATION OF PLATE.

—•—

# MOLLUSCA.

—•—

### PLATE XVII.

*Figs. 1, 1a.—Atys submalleata*, E. A. Smith, Ann. Mag. Nat. Hist., Ser. 7, Vol. XIV, July, 1904, p. 6.

*Figs. 2, 2a.—Cylichna andamanica*, E. A. Smith, *Op. cit.*, p. 6.

*Figs. 3, 3a.—Xylophaga indica*, E. A. Smith, *Op. cit.*, p. 7.

*Figs. 4, 4a, 4b, 4c.—Vesicomya indica*, E. A. Smith, *Op. cit.*, p. 9.

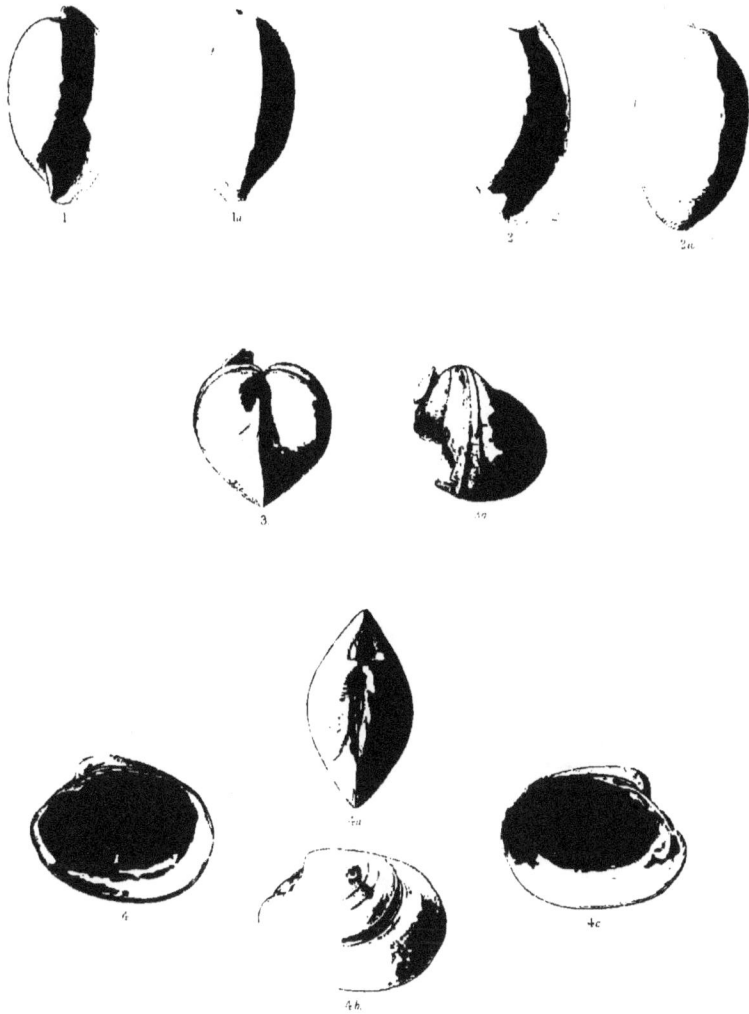

1 Atys suamalleata, E A Smith    2 Cylichna andamanica, E A Smith    3 Xylophaga nubea, E A Smith    4 Vesicomya indica, E A Smith

# EXPLANATION OF PLATE.

## MOLLUSCA.

### PLATE XVIII.

*Figs. 1, 1a.*—*Psammobia arakanenis,* E. A. Smith, Ann. Mag. Nat. Hist., Ser. 7, Vol. XIV, July, 1904, p. 10.

*Figs. 2, 2a.*—*Myrina indica,* E. A. Smith, *Op. cit.,* p. 11.

*Figs. 3, 3a.*—*Anatina andamanica,* E. A. Smith, *Op. cit.,* p. 8.

*Figs. 4, 4a, 4b.*—*Pecten alcocki,* E. A. Smith, *Op. cit.,* p. 13.

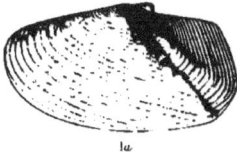

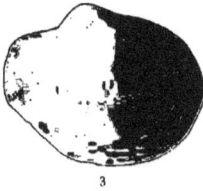

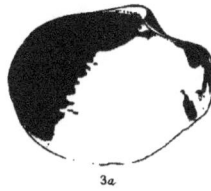

1 Psammobia arakanensis, E. A. Smith × 2   2  Myrina indica, E. A. Smith × 8   3. Anatina andamanica, E. A. Smith × 8   4. Pecten alcocki, E. A. Smith × 2½

$MoO - Pl. \, \text{V}, \, \text{Fig.} \ x, x \cdots xx$
1908

# EXPLANATION OF PLATE.

## PLATE XIX—MOLLUSCA.

*Figs. 1, 2.*—NATICA DIMIDIATA, E. A. Smith.

Natica dimidiata, *E. A. Smith, Ann. Mag. Nat. Hist.* (7), Vol. XVIII, p. 172 (1906).

*Figs. 3, 4.*—NATICA INCERTA, E. A. Smith.

Natica incerta, *E. A. Smith, op. cit.*, p. 173.

*Figs. 5, 6.*—SCAPHANDER VICINUS, E. A. Smith.

Scaphander vicinus, E. A. Smith, *Ann. Mag. Nat. Hist., tom. cit.*, p 248.

*Figs. 7, 8.*—SCAPHANDER ANDAMANICUS, E. A. Smith.

Scaphander andamanicus, *E. A. Smith, Ann. Mag. Nat. Hist.*, (6), Vol. XIV, p. 167, pl. IV, fig. 15 (1894); series 7, Vol. XIV, p. 5 (1904) and Vol. XVIII, p. 247 (1906).

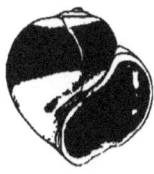

1.

5.

3.

2.

6.

4.

7.

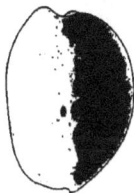

8.

1.2  Natica dimidiata, Smith × z.   5.6  Scaphander vicinus Smith × 1
3.4  Natica incerta Smith × 1       7.8  Scaphander andamaneus, Smith × 1.

# EXPLANATION OF PLATE.

## PLATE XX—MOLLUSCA.

*Figs. 1, 2.*—PLEUROTOMA (SURCULA) THISBE, E. A. Smith.
Pleurotoma thisbe, *E. A. Smith, Ann. Mag. Nat. Hist.* (7), Vol.
XVIII, p. 162 (1906).

*Figs. 3, 4.*—PLEUROTOMA CARINATA, Gray.
Pleurotoma carinata, Gray, *E. A. Smith, Ann. Mag. Nat. Hist.* (6),
Vol. XXIII, p. 368 (1896), and (7), Vol. XVIII, p. 160 (1906).

*Figs. 5, 6.*—ANCILLA ALCOCKII, E. A. Smith.
Ancilla alcockii, *E. A. Smith, Ann. Mag. Nat. Hist.* (7), Vol.
XVIII, p. 172 (1906).

*Figs. 7, 8.*—VESICOMYA BREVIS, E. A. Smith.
Vesicomya brevis, *E. A. Smith, op. cit*, p. 261.

5

2

7.

8.

6

3

4

EXPLANATION OF PLATE.

—•—

# PLATE XXI--MOLLUSCA.

—•—

*Figs. 1, 1a.*—PONTIOTHAUMA MINUS, E. A. Smith.
Pontiothauma minus, *E. A. Smith, Ann. Mag. Nat. Hist.* (7), Vol.
XVIII, p. 159 (1906).

*Figs. 2, 2a.*—PONTIOTHAUMA PACEI, E. A. Smith.
Pontiothauma pacei, *E. A. Smith, op. cit.*, p. 159.

*Figs. 3, 3a.*—PLEUROTOMA (SURCULA) NEREIS, E. A. Smith.
Pleurotoma (Surcula) nereis, *E. A. Smith, op. cit.*, p. 161.

*Figs. 4, 4a.*—PLEUROTOMA (SURCULA) AGALMA, E. A. Smith.
Pleurotoma (Surcula) agalma, *E. A. Smith, op. cit.*, p. 162.

*Figs. 5, 5a.*—PLEUROTOMA (BATHYTOMA) URANIA,
E. A. Smith.
Pleurotoma (Bathytoma) urania, *E. A. Smith, op. cit.*, p. 164.

*Figs. 6, 6a.*—MANARIA THURSTONI, E. A. Smith.
Manaria thurstoni, *E. A. Smith, op. cit.*, p. 167.

1      1a      2      2a

3      3a      4      4a

5      5a      6      6a

1, 1a.  Pontiothauma minus, Smith, × 2.  
2, 2a.  Pontiothauma pacoi, Smith.  
3, 3a.  Pleurotoma (Surcula) nereis, Smith, × 2.  

4, 4a.  Pleurotoma (Surcula) agulhas, Smith, × 4.  
5, 5a.  Pleurotoma (Bathytoma) urania, Smith, × 3.  
6, 6a.  Manaria thurstoni, Smith, × 2.

# EXPLANATION OF PLATE.

## PLATE XXII—MOLLUSCA.

***Figs. 1, 1a.***—NATICA SIMULANS, E. A. Smith.
Natica simulans, *E. A. Smith, Ann. Mag. Nat. Hist.* (7),
Vol. XVIII, p. 173 (1906).

***Figs. 2, 2a.***—COLUMBELLA SUAVIS, E. A. Smith.
Columbella suavis, *E. A. Smith, op. cit.*, p. 171.

***Figs. 3, 3a.***—TROPHON (BOREOTROPHON) PLANI-
SPINA, E. A. Smith.
Trophon (Boreotrophon) planispina, *E. A. Smith, op. cit.*, p. 168.

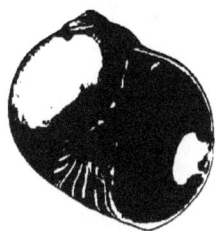

1

1a

2

2a

3

3a

EXPLANATION OF PLATE.

## PLATE XXIII—MOLLUSCA.

*Figs. 1, 1a.*—NUCULA (ACILA) GRANULATA, E. A. Smith.
Nucula (Acila) granulata, *E. A. Smith, Ann. Mag Nat. Hist.*
(7), Vol. XVIII, p. 251 (1906).

*Figs. 2, 2a.*—DENTALIUM CORNU-BOVIS, E. A Smith
Dentalium cornu-bovis, *E. A. Smith, op. cit.*, p. 249.

N.B.—Figs. 2 and 2a with Symbiotic Actinian.

*Fig. 3.*—DENTALIUM SUBCURVATUM, E. A. Smith.
Dentalium subcurvatum, *E. A. Smith, op. cit.*, p. 251.

*Figs. 4, 4a.*—LEPIDOPLEURUS ANDAMANICUS, E. A. Smith.
Lepidopleurus andamanicus, *E. A. Smith, op. cit.*, p. 251.

1

1a

2

2a

3

4

4a

Half tone.

Survey of India Offices, Calcutta. 1909

1, 1a.  Nucula (Acila) granulata, Smith, × 4.
2, 2a.  Dentalium cornu-bovis, Smith, × 1½.

3.  Dentalium subcurvatum, Smith, × 1½.
4, 4a.  Lepidopleurus andamanensis, Smith, × 5.